# ENGINEERING
# NOTEBOOK

| NAME | |
|---|---|
| SIGNATURE | DATE |
| DATE ISSUED | BY |
| PHONE | EMAIL |
| COMPANY | |
| DEPARTMENT | |
| ADDRESS | |
| | |

| CITY | STATE | ZIP |
|---|---|---|

NOTES

# TABLE OF CONTENTS

| PAGE | SUBJECT | DATE |
|:---:|---|---|
| 1 | | |
| 2 | | |
| 3 | | |
| 4 | | |
| 5 | | |
| 6 | | |
| 7 | | |
| 8 | | |
| 9 | | |
| 10 | | |
| 11 | | |
| 12 | | |
| 13 | | |
| 14 | | |
| 15 | | |
| 16 | | |
| 17 | | |
| 18 | | |
| 19 | | |
| 20 | | |
| 21 | | |
| 22 | | |
| 23 | | |
| 24 | | |
| 25 | | |
| 26 | | |
| 27 | | |
| 28 | | |
| 29 | | |
| 30 | | |

# TABLE OF CONTENTS

| PAGE | SUBJECT | DATE |
|------|---------|------|
| 31 | | |
| 32 | | |
| 33 | | |
| 34 | | |
| 35 | | |
| 36 | | |
| 37 | | |
| 38 | | |
| 39 | | |
| 40 | | |
| 41 | | |
| 42 | | |
| 43 | | |
| 44 | | |
| 45 | | |
| 46 | | |
| 47 | | |
| 48 | | |
| 49 | | |
| 50 | | |
| 51 | | |
| 52 | | |
| 53 | | |
| 54 | | |
| 55 | | |
| 56 | | |
| 57 | | |
| 58 | | |
| 59 | | |
| 60 | | |

# TABLE OF CONTENTS

| PAGE | SUBJECT | DATE |
|:----:|:--------|:----:|
| 61 | | |
| 62 | | |
| 63 | | |
| 64 | | |
| 65 | | |
| 66 | | |
| 67 | | |
| 68 | | |
| 69 | | |
| 70 | | |
| 71 | | |
| 72 | | |
| 73 | | |
| 74 | | |
| 75 | | |
| 76 | | |
| 77 | | |
| 78 | | |
| 79 | | |
| 80 | | |
| 81 | | |
| 82 | | |
| 83 | | |
| 84 | | |
| 85 | | |
| 86 | | |
| 87 | | |
| 88 | | |
| 89 | | |
| 90 | | |

# TABLE OF CONTENTS

| PAGE | SUBJECT | DATE |
|------|---------|------|
| 91 | | |
| 92 | | |
| 93 | | |
| 94 | | |
| 95 | | |
| 96 | | |
| 97 | | |
| 98 | | |
| 99 | | |
| 100 | | |
| 101 | | |
| 102 | | |
| 103 | | |
| 104 | | |
| 105 | | |
| 106 | | |
| 107 | | |
| 108 | | |
| 109 | | |
| 110 | | |
| 111 | | |
| 112 | | |
| 113 | | |
| 114 | | |
| 115 | | |
| 116 | | |
| 117 | | |
| 118 | | |
| 119 | | |
| 120 | | |

**TITLE**

**PROJECT**

*Continued from Page*

1

5

10

15

20

25

30

35

*Continued to Page*

SIGNATURE

DATE

DISCLOSED TO AND UNDERSTOOD BY

DATE

**PROPRIETARY INFORMATION**

*Continued from Page*

**2**

5

10

15

20

25

30

35

*Continued to Page*

SIGNATURE

DATE

DISCLOSED TO AND UNDERSTOOD BY

DATE

**PROPRIETARY INFORMATION**

*Continued from Page*

**3**

5

10

15

20

25

30

35

*Continued to Page*

*Continued from Page*

**4**

5

10

15

20

25

30

35

*Continued to Page*

SIGNATURE

DATE

DISCLOSED TO AND UNDERSTOOD BY

DATE

**PROPRIETARY INFORMATION**

**TITLE**                                           **PROJECT**

*Continued from Page*

5

10

15

20

25

30

35

*Continued to Page*

SIGNATURE                                                          DATE

DISCLOSED TO AND UNDERSTOOD BY              DATE              **PROPRIETARY INFORMATION**

**TITLE**  **PROJECT**

*Continued from Page*

**6**

5

10

15

20

25

30

35

*Continued to Page*

SIGNATURE

DATE

DISCLOSED TO AND UNDERSTOOD BY

DATE

**PROPRIETARY INFORMATION**

**TITLE**

**PROJECT**

7

*Continued from Page*

5

10

15

20

25

30

35

*Continued to Page*

SIGNATURE

DATE

DISCLOSED TO AND UNDERSTOOD BY

DATE

**PROPRIETARY INFORMATION**

*Continued from Page*

**8**

5

10

15

20

25

30

35

*Continued to Page*

SIGNATURE

DATE

DISCLOSED TO AND UNDERSTOOD BY

DATE

Continued from Page

5

10

15

20

25

30

35

Continued to Page

SIGNATURE

DATE

DISCLOSED TO AND UNDERSTOOD BY

DATE

**PROPRIETARY INFORMATION**

*Continued from Page*

**10**

5

10

15

20

25

30

35

*Continued to Page*

SIGNATURE

DATE

DISCLOSED TO AND UNDERSTOOD BY

DATE

**PROPRIETARY INFORMATION**

*Continued from Page*

5

10

15

20

25

30

35

*Continued to Page*

SIGNATURE

DATE

DISCLOSED TO AND UNDERSTOOD BY

DATE

**PROPRIETARY INFORMATION**

**TITLE**　　　　　　　　　　**PROJECT**

Continlued from Page

5

10

15

20

25

30

35

Continued to Page

SIGNATURE　　　　　　　　　　　　　　　　　DATE

DISCLOSED TO AND UNDERSTOOD BY　　　　　　DATE

**PROPRIETARY INFORMATION**

**TITLE**

**PROJECT**

Continued from Page

**13**

5

10

15

20

25

30

35

Continued to Page

SIGNATURE

DATE

DISCLOSED TO AND UNDERSTOOD BY

DATE

**PROPRIETARY INFORMATION**

*Continued from Page*

5

10

15

20

25

30

35

*Continued to Page*

SIGNATURE

DATE

DISCLOSED TO AND UNDERSTOOD BY

DATE

**PROPRIETARY INFORMATION**

*Continued from Page*

5

10

15

20

25

30

35

*Continued to Page*

SIGNATURE

DATE

DISCLOSED TO AND UNDERSTOOD BY

DATE

**PROPRIETARY INFORMATION**

*Continued from Page*

5

10

15

20

25

30

35

*Continued to Page*

SIGNATURE

DATE

DISCLOSED TO AND UNDERSTOOD BY

DATE

**PROPRIETARY INFORMATION**

*Continued from Page*

**17**

5

10

15

20

25

30

35

*Continued to Page*

SIGNATURE

DATE

DISCLOSED TO AND UNDERSTOOD BY

DATE

**PROPRIETARY INFORMATION**

*Continued from Page*

5

10

15

20

25

30

35

*Continued to Page*

SIGNATURE

DATE

DISCLOSED TO AND UNDERSTOOD BY

DATE

*Continued from Page*

5

10

15

20

25

30

35

*Continued to Page*

SIGNATURE                                       DATE

DISCLOSED TO AND UNDERSTOOD BY            DATE                 **PROPRIETARY INFORMATION**

*Continued from Page*

**20**

5

10

15

20

25

30

35

*Continued to Page*

SIGNATURE

DATE

DISCLOSED TO AND UNDERSTOOD BY

DATE

**PROPRIETARY INFORMATION**

TITLE

PROJECT

*Continued from Page*

**21**

5

10

15

20

25

30

35

*Continued to Page*

SIGNATURE

DATE

DISCLOSED TO AND UNDERSTOOD BY

DATE

**PROPRIETARY INFORMATION**

**22**

5

10

15

20

25

30

35

*Continued to Page*

SIGNATURE

DATE

DISCLOSED TO AND UNDERSTOOD BY

DATE

**PROPRIETARY INFORMATION**

**TITLE**                           **PROJECT**

*Continued from Page*

**23**

5

10

15

20

25

30

35

*Continued to Page*

SIGNATURE                                              DATE

DISCLOSED TO AND UNDERSTOOD BY          DATE          **PROPRIETARY INFORMATION**

*Continued from Page*

5

10

15

20

25

30

35

*Continued to Page*

SIGNATURE

DATE

DISCLOSED TO AND UNDERSTOOD BY

DATE

**PROPRIETARY INFORMATION**

*Continued from Page*

5

10

15

20

25

30

35

*Continued to Page*

SIGNATURE

DATE

DISCLOSED TO AND UNDERSTOOD BY

DATE

**PROPRIETARY INFORMATION**

*Continued from Page*

**26**

5

10

15

20

25

30

35

*Continued to Page*

SIGNATURE

DATE

DISCLOSED TO AND UNDERSTOOD BY

DATE

**PROPRIETARY INFORMATION**

**27**

5

10

15

20

25

30

35

*Continued to Page*

SIGNATURE

DATE

DISCLOSED TO AND UNDERSTOOD BY

DATE

**PROPRIETARY INFORMATION**

*Continued from Page*

*Continued to Page*

SIGNATURE

DATE

DISCLOSED TO AND UNDERSTOOD BY

DATE

**TITLE**

*Continued from Page*

**29**

5

10

15

20

25

30

35

*Continued to Page*

SIGNATURE

DATE

DISCLOSED TO AND UNDERSTOOD BY

DATE

**PROPRIETARY INFORMATION**

*Continued from Page*

**30**

5

10

15

20

25

30

35

*Continued to Page*

SIGNATURE

DATE

DISCLOSED TO AND UNDERSTOOD BY

DATE

**PROPRIETARY INFORMATION**

**TITLE**

**PROJECT**

*Continued from Page*

**31**

5

10

15

20

25

30

35

*Continued to Page*

SIGNATURE

DATE

DISCLOSED TO AND UNDERSTOOD BY

DATE

**PROPRIETARY INFORMATION**

*Continued from Page*

5

10

15

20

25

30

35

*Continued to Page*

SIGNATURE

DATE

DISCLOSED TO AND UNDERSTOOD BY

DATE

**PROPRIETARY INFORMATION**

*Continued from Page*

**33**

5

10

15

20

25

30

35

*Continued to Page*

SIGNATURE　　　　　　　　　　　　　　　　　　DATE

*Continued from Page*

*Continued to Page*

SIGNATURE

DATE

DISCLOSED TO AND UNDERSTOOD BY

DATE

*Continued from Page*

*Continued to Page*

SIGNATURE

DATE

DISCLOSED TO AND UNDERSTOOD BY

DATE

*Continued from Page*

**36**

5

10

15

20

25

30

35

*Continued to Page*

SIGNATURE

DATE

DISCLOSED TO AND UNDERSTOOD BY

DATE

**PROPRIETARY INFORMATION**

*Continued from Page*

*Continued to Page*

SIGNATURE

DATE

DISCLOSED TO AND UNDERSTOOD BY

DATE

*Continued from Page*

5

10

15

20

25

30

35

*Continued to Page*

SIGNATURE

DATE

DISCLOSED TO AND UNDERSTOOD BY

DATE

*Continued from Page*

5

10

15

20

25

30

35

*Continued to Page*

SIGNATURE

DATE

DISCLOSED TO AND UNDERSTOOD BY

DATE

**PROPRIETARY INFORMATION**

*Continued from Page*

5

10

15

20

25

30

35

*Continued to Page*

SIGNATURE

DATE

DISCLOSED TO AND UNDERSTOOD BY

DATE

**PROPRIETARY INFORMATION**

*Continued from Page*

41

5

10

15

20

25

30

35

*Continued to Page*

SIGNATURE

DATE

DISCLOSED TO AND UNDERSTOOD BY

DATE

**PROPRIETARY INFORMATION**

*Continued from Page*

*Continued to Page*

SIGNATURE

DATE

DISCLOSED TO AND UNDERSTOOD BY

DATE

*Continued from Page*

5

10

15

20

25

30

35

*Continued to Page*

SIGNATURE

DATE

DISCLOSED TO AND UNDERSTOOD BY

DATE

**PROPRIETARY INFORMATION**

*Continued from Page*

5

10

15

20

25

30

35

*Continued to Page*

SIGNATURE

DATE

DISCLOSED TO AND UNDERSTOOD BY

DATE

**PROPRIETARY INFORMATION**

*Continued from Page*

5

10

15

20

25

30

35

*Continued to Page*

SIGNATURE          DATE

DISCLOSED TO AND UNDERSTOOD BY      DATE

**PROPRIETARY INFORMATION**

Continued from Page

5

10

15

20

25

30

35

Continued to Page

SIGNATURE

DATE

DISCLOSED TO AND UNDERSTOOD BY

DATE

*Continued from Page*

*Continued to Page*

SIGNATURE

DATE

DISCLOSED TO AND UNDERSTOOD BY

DATE

Continued from Page

**48**

5

10

15

20

25

30

35

Continued to Page

SIGNATURE

DATE

DISCLOSED TO AND UNDERSTOOD BY

DATE

**PROPRIETARY INFORMATION**

*Continued from Page*

5

10

15

20

25

30

35

*Continued to Page*

SIGNATURE                                                    DATE

DISCLOSED TO AND UNDERSTOOD BY          DATE          **PROPRIETARY INFORMATION**

*Continued from Page*

*Continued to Page*

SIGNATURE

DATE

DISCLOSED TO AND UNDERSTOOD BY

DATE

**PROPRIETARY INFORMATION**

**TITLE**

**PROJECT**

*Continued from Page*

**51**

5

10

15

20

25

30

35

*Continued to Page*

SIGNATURE

DATE

DISCLOSED TO AND UNDERSTOOD BY

DATE

**PROPRIETARY INFORMATION**

*Continued from Page*

*Continued to Page*

SIGNATURE

DATE

DISCLOSED TO AND UNDERSTOOD BY

DATE

**PROPRIETARY INFORMATION**

*Continued from Page*

**53**

5

10

15

20

25

30

35

*Continued to Page*

SIGNATURE

DATE

DISCLOSED TO AND UNDERSTOOD BY

DATE

**PROPRIETARY INFORMATION**

*Continued from Page*

**54**

*Continued to Page*

SIGNATURE

DATE

DISCLOSED TO AND UNDERSTOOD BY

DATE

**PROPRIETARY INFORMATION**

*Continued from Page*

**55**

5

10

15

20

25

30

35

*Continued to Page*

SIGNATURE

DATE

DISCLOSED TO AND UNDERSTOOD BY

DATE

**PROPRIETARY INFORMATION**

Continued from Page

**56**

5

10

15

20

25

30

35

Continued to Page

SIGNATURE

DATE

DISCLOSED TO AND UNDERSTOOD BY

DATE

**PROPRIETARY INFORMATION**

*Continued from Page*

5

10

15

20

25

30

35

*Continued to Page*

SIGNATURE

DATE

DISCLOSED TO AND UNDERSTOOD BY

DATE

**PROPRIETARY INFORMATION**

*Continued from Page*

**58**

5

10

15

20

25

30

35

*Continued to Page*

SIGNATURE

DATE

DISCLOSED TO AND UNDERSTOOD BY

DATE

**PROPRIETARY INFORMATION**

*Continued from Page*

**59**

5

10

15

20

25

30

35

*Continued to Page*

SIGNATURE

DATE

DISCLOSED TO AND UNDERSTOOD BY

DATE

**PROPRIETARY INFORMATION**

*Continued from Page*

5

10

15

20

25

30

35

*Continued to Page*

SIGNATURE

DATE

DISCLOSED TO AND UNDERSTOOD BY

DATE

**PROPRIETARY INFORMATION**

Continued from Page

5

10

15

20

25

30

35

Continued to Page

SIGNATURE

DATE

DISCLOSED TO AND UNDERSTOOD BY

DATE

**PROPRIETARY INFORMATION**

*Continued from Page*

**62**

5

10

15

20

25

30

35

*Continued to Page*

SIGNATURE

DATE

DISCLOSED TO AND UNDERSTOOD BY

DATE

**PROPRIETARY INFORMATION**

*Continued from Page*

5

10

15

20

25

30

35

*Continued to Page*

SIGNATURE

DATE

DISCLOSED TO AND UNDERSTOOD BY

DATE

*Continued from Page*

5

10

15

20

25

30

35

*Continued to Page*

SIGNATURE

DATE

DISCLOSED TO AND UNDERSTOOD BY

DATE

**PROPRIETARY INFORMATION**

*Continued from Page*

5

10

15

20

25

30

35

*Continued to Page*

SIGNATURE

DATE

DISCLOSED TO AND UNDERSTOOD BY

DATE

**PROPRIETARY INFORMATION**

*Continued from Page*

5

10

15

20

25

30

35

*Continued to Page*

SIGNATURE

DATE

DISCLOSED TO AND UNDERSTOOD BY

DATE

*Continued from Page*

5

10

15

20

25

30

35

*Continued to Page*

SIGNATURE

DATE

DISCLOSED TO AND UNDERSTOOD BY

DATE

**PROPRIETARY INFORMATION**

*Continued from Page*

**68**

5

10

15

20

25

30

35

*Continued to Page*

SIGNATURE

DATE

DISCLOSED TO AND UNDERSTOOD BY

DATE

**PROPRIETARY INFORMATION**

*Continued from Page*

5

10

15

20

25

30

35

*Continued to Page*

*Continued from Page*

5

10

15

20

25

30

35

*Continued to Page*

SIGNATURE

DATE

DISCLOSED TO AND UNDERSTOOD BY

DATE

**PROPRIETARY INFORMATION**

**TITLE**　　　　　　　　**PROJECT**

*Continued from Page*

**71**

*Continued to Page*

SIGNATURE

DATE

DISCLOSED TO AND UNDERSTOOD BY

DATE

**PROPRIETARY INFORMATION**

*Continued from Page*

**72**

5

10

15

20

25

30

35

*Continued to Page*

SIGNATURE

DATE

DISCLOSED TO AND UNDERSTOOD BY

DATE

**PROPRIETARY INFORMATION**

*Continued from Page*

**73**

5

10

15

20

25

30

35

*Continued to Page*

SIGNATURE

DATE

DISCLOSED TO AND UNDERSTOOD BY

DATE

**PROPRIETARY INFORMATION**

*Continued from Page*

5

10

15

20

25

30

35

*Continued to Page*

SIGNATURE                    DATE

DISCLOSED TO AND UNDERSTOOD BY          DATE          **PROPRIETARY INFORMATION**

*Continued from Page*

5

10

15

20

25

30

35

*Continued to Page*

SIGNATURE

DATE

DISCLOSED TO AND UNDERSTOOD BY

DATE

**PROPRIETARY INFORMATION**

Continued from Page

5

10

15

20

25

30

35

Continued to Page

SIGNATURE

DATE

DISCLOSED TO AND UNDERSTOOD BY

DATE

**PROPRIETARY INFORMATION**

*Continued from Page*

5

10

15

20

25

30

35

*Continued to Page*

SIGNATURE

DATE

DISCLOSED TO AND UNDERSTOOD BY

DATE

**PROPRIETARY INFORMATION**

**TITLE**　　　　　　　　　　　　　**PROJECT**

*Continued from Page*

**78**

5

10

15

20

25

30

35

*Continued to Page*

SIGNATURE

DATE

DISCLOSED TO AND UNDERSTOOD BY

DATE

**PROPRIETARY INFORMATION**

*Continued from Page*

5

10

15

20

25

30

35

*Continued to Page*

SIGNATURE                                 DATE

DISCLOSED TO AND UNDERSTOOD BY           DATE           **PROPRIETARY INFORMATION**

Continued from Page

Continued to Page

SIGNATURE

DATE

DISCLOSED TO AND UNDERSTOOD BY

DATE

**TITLE**

**PROJECT**

Continued from Page

**81**

5

10

15

20

25

30

35

Continued to Page

SIGNATURE

DATE

DISCLOSED TO AND UNDERSTOOD BY

DATE

**PROPRIETARY INFORMATION**

*Continued from Page*

5

10

15

20

25

30

35

*Continued to Page*

SIGNATURE

DATE

DISCLOSED TO AND UNDERSTOOD BY

DATE

**PROPRIETARY INFORMATION**

*Continued from Page*

5

10

15

20

25

30

35

*Continued to Page*

SIGNATURE

DATE

DISCLOSED TO AND UNDERSTOOD BY

DATE

**PROPRIETARY INFORMATION**

*Continued from Page*

5

10

15

20

25

30

35

*Continued to Page*

SIGNATURE

DATE

DISCLOSED TO AND UNDERSTOOD BY

DATE

*Continued from Page*

5

10

15

20

25

30

35

*Continued to Page*

SIGNATURE                                                    DATE

*Continued from Page*

5

10

15

20

25

30

35

*Continued to Page*

SIGNATURE

DATE

DISCLOSED TO AND UNDERSTOOD BY

DATE

**PROPRIETARY INFORMATION**

*Continued from Page*

5

10

15

20

25

30

35

*Continued to Page*

SIGNATURE

DATE

DISCLOSED TO AND UNDERSTOOD BY

DATE

**PROPRIETARY INFORMATION**

*Continued from Page*

5

10

15

20

25

30

35

*Continued to Page*

SIGNATURE

DATE

DISCLOSED TO AND UNDERSTOOD BY

DATE

**PROPRIETARY INFORMATION**

*Continued from Page*

**89**

5

10

15

20

25

30

35

*Continued to Page*

SIGNATURE

DATE

DISCLOSED TO AND UNDERSTOOD BY

DATE

**PROPRIETARY INFORMATION**

*Continued from Page*

**90**

5

10

15

20

25

30

35

*Continued to Page*

SIGNATURE

DATE

DISCLOSED TO AND UNDERSTOOD BY

DATE

**PROPRIETARY INFORMATION**

*Continued from Page*

5

10

15

20

25

30

35

*Continued to Page*

SIGNATURE    DATE

DISCLOSED TO AND UNDERSTOOD BY    DATE    **PROPRIETARY INFORMATION**

*Continued from Page*

5

10

15

20

25

30

35

*Continued to Page*

SIGNATURE

DATE

DISCLOSED TO AND UNDERSTOOD BY

DATE

**PROPRIETARY INFORMATION**

*Continued from Page*

5

10

15

20

25

30

35

*Continued to Page*

SIGNATURE

DATE

DISCLOSED TO AND UNDERSTOOD BY

DATE

**PROPRIETARY INFORMATION**

Continued from Page

5

10

15

20

25

30

35

Continued to Page

SIGNATURE

DATE

DISCLOSED TO AND UNDERSTOOD BY

DATE

**PROPRIETARY INFORMATION**

*Continued from Page*

*Continued to Page*

SIGNATURE

DATE

DISCLOSED TO AND UNDERSTOOD BY

DATE

*Continued from Page*

5

10

15

20

25

30

35

*Continued to Page*

SIGNATURE

DATE

DISCLOSED TO AND UNDERSTOOD BY

DATE

**PROPRIETARY INFORMATION**

5

10

15

20

25

30

35

*Continued to Page*

SIGNATURE

DATE

DISCLOSED TO AND UNDERSTOOD BY

DATE

**PROPRIETARY INFORMATION**

*Continued from Page*

5

10

15

20

25

30

35

*Continued to Page*

SIGNATURE

DATE

DISCLOSED TO AND UNDERSTOOD BY

DATE

**PROPRIETARY INFORMATION**

**TITLE**

**PROJECT**

*Continued from Page*

99

5

10

15

20

25

30

35

*Continued to Page*

SIGNATURE

DATE

DISCLOSED TO AND UNDERSTOOD BY

DATE

**PROPRIETARY INFORMATION**

**TITLE**

**PROJECT**

Continued from Page

100

5

10

15

20

25

30

35

Continued to Page

SIGNATURE

DATE

DISCLOSED TO AND UNDERSTOOD BY

DATE

**PROPRIETARY INFORMATION**

*Continued from Page*

5

10

15

20

25

30

35

*Continued to Page*

SIGNATURE

DATE

DISCLOSED TO AND UNDERSTOOD BY

DATE

**PROPRIETARY INFORMATION**

*Continued from Page*

5

10

15

20

25

30

35

*Continued to Page*

SIGNATURE

DATE

DISCLOSED TO AND UNDERSTOOD BY

DATE

**PROPRIETARY INFORMATION**

*Continued from Page*

5

10

15

20

25

30

35

*Continued to Page*

SIGNATURE

DATE

DISCLOSED TO AND UNDERSTOOD BY

DATE

**PROPRIETARY INFORMATION**

Continued from Page

5

10

15

20

25

30

35

Continued to Page

SIGNATURE

DATE

DISCLOSED TO AND UNDERSTOOD BY

DATE

**PROPRIETARY INFORMATION**

Continued from Page

5

10

15

20

25

30

35

Continued to Page

SIGNATURE

DATE

DISCLOSED TO AND UNDERSTOOD BY

DATE

**PROPRIETARY INFORMATION**

Continued from Page

5

10

15

20

25

30

35

Continued to Page

SIGNATURE

DATE

DISCLOSED TO AND UNDERSTOOD BY

DATE

**PROPRIETARY INFORMATION**

*Continued from Page*

5

10

15

20

25

30

35

*Continued to Page*

SIGNATURE

DATE

DISCLOSED TO AND UNDERSTOOD BY

DATE

**PROPRIETARY INFORMATION**

*Continued from Page*

*Continued to Page*

SIGNATURE

DATE

DISCLOSED TO AND UNDERSTOOD BY

DATE

*Continued from Page*

5

10

15

20

25

30

35

*Continued to Page*

SIGNATURE

DATE

DISCLOSED TO AND UNDERSTOOD BY

DATE

**PROPRIETARY INFORMATION**

*Continued from Page*

5

10

15

20

25

30

35

*Continued to Page*

SIGNATURE

DATE

DISCLOSED TO AND UNDERSTOOD BY

DATE

**PROPRIETARY INFORMATION**

*Continued from Page*

**111**

5

10

15

20

25

30

35

*Continued to Page*

SIGNATURE

DATE

DISCLOSED TO AND UNDERSTOOD BY

DATE

**PROPRIETARY INFORMATION**

*Continued from Page*

5

10

15

20

25

30

35

*Continued to Page*

SIGNATURE

DATE

DISCLOSED TO AND UNDERSTOOD BY

DATE

*Continued from Page*

5

10

15

20

25

30

35

*Continued to Page*

SIGNATURE

DATE

DISCLOSED TO AND UNDERSTOOD BY

DATE

**PROPRIETARY INFORMATION**

*Continued from Page*

5

10

15

20

25

30

35

*Continued to Page*

SIGNATURE

DATE

DISCLOSED TO AND UNDERSTOOD BY

DATE

**PROPRIETARY INFORMATION**

*Continued from Page*

5

10

15

20

25

30

35

*Continued to Page*

SIGNATURE

DATE

DISCLOSED TO AND UNDERSTOOD BY

DATE

**PROPRIETARY INFORMATION**

*Continued from Page*

*Continued to Page*

SIGNATURE

DATE

DISCLOSED TO AND UNDERSTOOD BY

DATE

**PROPRIETARY INFORMATION**

*Continued from Page*

5

10

15

20

25

30

35

*Continued to Page*

SIGNATURE

DATE

DISCLOSED TO AND UNDERSTOOD BY

DATE

**PROPRIETARY INFORMATION**

**TITLE**

**PROJECT**

*Continued from Page*

**118**

5

10

15

20

25

30

35

*Continued to Page*

SIGNATURE

DATE

DISCLOSED TO AND UNDERSTOOD BY

DATE

**PROPRIETARY INFORMATION**

**TITLE**

**PROJECT**

Continued from Page

119

5

10

15

20

25

30

35

Continued to Page

SIGNATURE

DATE

DISCLOSED TO AND UNDERSTOOD BY

DATE

**PROPRIETARY INFORMATION**

*Continued from Page*

5

10

15

20

25

30

35

*Continued to Page*

SIGNATURE

DATE

DISCLOSED TO AND UNDERSTOOD BY

DATE

**PROPRIETARY INFORMATION**

Made in the USA
Monee, IL
28 November 2023

47645139R00070